Creation

A Scientific Perspective

R. Dale Potter

Introduction

I do not claim to have any special insight into the Word of God or into science. The Apostle Paul stated in Philippians 4:8-9: *Finally, brethren, whatsoever things are true, whatsoever things are honest, whatsoever things are just, whatsoever things are pure, whatsoever things are lovely, whatsoever things are of good report; if there be any virtue, and if there be any praise, think on these things. Those things, which ye have both learned, and received, and heard, and seen in me, do: and the God of peace shall be with you.*

My desire is that everything that I state in these writings upholds this great statement in its entirety. I do understand that I, and only I, will be accountable to our Lord and Savior for everything that I teach; therefore, I wish to magnify and glorify Him and use His Word as a standard that will prove my own words, lest I be weighed in the balances and found wanting.

Another scripture that goes hand in hand with this one is Rom. 12:1-2, where Paul is admonishing us not to be conformed to this world, but to be transformed by the renewing of our minds. We should not think like the world anymore, but like Christ.

Preface

The objective of this book is to show how true science correlates with the Word of God in creation. I am not a scientist, though I do follow science to some detail. Notwithstanding, I also believe that science and math are integral parts of the creation of God. Therefore, I go into detail explaining the symmetry of the universe from a biblical point of view.

I also try to show how theories such as the Big Bang theory do not hold water, so to speak, when confronted with true biblical doctrine. I am a certified electronics technician/troubleshooter holding a degree in Electronics Engineering Technology. However, I have been retired on disability for many years. I am also an ordained deacon and singer by the Grace of God.

It seems that today, the escalation of so many theories has grown until it is a task in itself to bring any validity to the real concepts of physics and science. I have heard things on YouTube that make my head swim with the ignorance of it all. Even though I wrote this book many years ago, it is as a result of these false concepts, as well as my pastor urging me, that I have spent much time in prayer and decided that it would be God's will for me to seek the publishing of this literature.

I do not claim to be a writer, but I feel as though God has inspired these writings, and I seek only to glorify Him and hopefully edify and bring His wonderful Word to light. Last of all, the proceeds from this book will go toward helping the needy and supporting missions that I contact. I will, however, keep enough royalties to cover the cost of publishing. May God richly bless you all with His bountiful Mercy, Grace, and Love.

In the Beginning

In the beginning God created the heaven and the earth. And the earth was without form and void; and darkness was upon the face of the deep. And the Spirit of God moved upon the face of the waters (Gen. 1:1-2).

The first two verses in the Bible are the most dynamic scriptures concerning how the Universe was formed and created. These are also the most controversial verses concerning the Creation of God. In my studies, I have come to a very interesting conclusion: God has never disproved science, which is true science, but science, as a whole, has gone to great lengths to try and disprove God's very existence. However, the only thing that science has done in the end is to further substantiate the Bible's validity. There is a sameness, or homogeneity, of the Universe that compels every creature to believe that God is the Creator.

The world's great physicist, Stephen Hawking, has come to the conclusion that the sum of everything in the Universe is NOTHING. What he is trying to relate is that the Universe was created from nothing, and there is no God who created it because it came from nothing. Well, this is a very interesting concept because God has already said, and preachers have preached for as long as I can remember, that God created the Universe from NOTHING. In fact, the only way that anything can be created is from nothing. That is why there is only one creation, and God took nothing and made something out of it; therefore, everything else in the Universe was made from what God created.

Many years ago, in electronics, I was taught Kirchhoff's Voltage Law, which states: *"the sum of the voltages at each node around a loop equals zero."* This is the same statement that Hawking made concerning the Universe, but it is more applicable. Think about it for just a moment in a mathematical sense. Note how Kirchhoff ended his statement by subtracting all the voltages in the loop from the source to come up with zero volts. This means that energy cannot be created or destroyed, but can only change states. Our great God of the Universe only made "X" amount of this stuff, and that is all there is.

So, we conclude that Kirchhoff rightly used the statement, and the theory is logical and mathematically true. But Hawking has a flawed answer to his use of the statement. Why? Because he is subtracting everything in the Universe from the whole of the Universe, and not the Source. Realizing that the Universe was created from nothing, because God said that it was, the only way that the theory can hold water is if the Source of the Universe is equal to something besides zero. Obviously, the Universe does exist, and the source IS zero, as far as elements are concerned.

The statement Hawking made is illogical because everything in the Universe subtracted from the source, zero, does not equal zero, but rather the negative of the sum. Logically, this statement does not hold up under scrutiny, and the whole theory falls apart because you must have a source that is equal to the Universe and not zero for the calculation to make sense.

I apologize to my readers for playing these word and number games, but so many people are drawn into misinterpretations simply because of the word-and-number-game theories that some well-educated scientists fling at them. When this happens, all we have to do is apply the basic rules of mathematics, which have never been known to fail, and see how it stands up against them.

Plato once said that *"the beauty of style, and grace, and harmony, and good rhythm, depend on simplicity."* God has given us His Word that we might learn, and He gave us logic that we might discern. Logic is the theory of reasoning, and discernment is the ability to understand something that is not clear. God said: *Come now, let us reason together…* (Isa. 1:18). It seems logical to me that science would progress much faster if scientists would put forth more effort to understand God's logic instead of trying to prove Him nonexistent.

Notwithstanding, it is much easier for me to believe that the sum of everything in the Universe equals a BIG GOD. Notwithstanding, there is no way that God can be formulated because He is infinite. Nothing in heaven or earth or eternity can exist without God, and that is why I used this analogy. God has stated throughout the entire Bible that we must only accept what He says by faith.

Blessed are they that believe through faith without having proof, but even so, it is incredible to think that scientific geniuses have spent their entire lives investigating and formulating the Universe and yet make grammar-school mistakes like this because they refuse to acknowledge the true source. My objective here is not to prove or disprove anything, but to correlate what is true science to the living Word of God. It makes no difference what we learn in a classroom environment; if it is factual, then it will be validated by the Word of God. The important thing to know is that a "Big God" did it, and His Word has always rung true all through the ages of time and even before time began.

With that said, I wish to relate that what I am saying in these writings is based upon fact, but it is still my theory of Creation. There are no errors in the Bible because it is absolute truth. Be that as it may, the God of the Universe does not need me to prove anything. My job is to state biblical facts that will back up what I say from a scientific point of view, and if there is anything that I say that isn't correct, then it is I who am wrong and not the Word of God.

In mathematics, there is a symbol that indicates an absolute value. If I say that an expression is equal to "A," then the answer may vary by some plus or minus amount, but if I state that an expression is equal to "|A|," then it is the absolute value and cannot vary. Therefore, God's Word = |TRUTH|; not a plus amount or a minus amount, because either would put the truth of God in error.

The place to start concerning biblical truth is at the beginning. The first four words of the Bible confirm this. *"In the beginning God"* is not when time began. Keep in mind that eternity has no beginning and no end, IT JUST IS; trying to place a time element or a form on eternity is where mankind has brought much confusion into the understanding of God and His logic and |Truth|.

In the beginning is not man's conceptual understanding of time; it only indicates an event. It simply states that this point inside the realm of eternity is where God decided to do a work. This is where God surveyed an area of space and saw nothing and, therefore, decided to create something that His eyes could behold and call *"Good."* It is where Creation began and not where time began. It is not a point in time, but rather a point in eternity.

Note also another very important concept: the Bible says that God created the *"heaven"* and the *"earth."* God is relating here that, of all the Universes that God has created throughout eternity, assuming that there were others, He is now concentrating fully on this one; not that He has stopped concentrating fully on the others as well, but that every one of them has His undivided attention. It is like a loving lady who wants her home to look astounding and beautiful. She is aware of each piece of furniture but is concentrating on placing a new addition in her home.

God is not only in heaven, but rather *"the Heavens."* We think of God as being in heaven and earth, and this is not a wrong analogy, but He also created other heavens and other earths, and He is in all of them because they are all a part of eternity. Isn't it amazing how big God really is, but yet He can also fit within the hearts of men?

Science and quantum mechanics have proven that if a flower is moved by a gentle breeze here on earth, then it is felt throughout the Universe. This is not a poetic point of view; it is a scientific fact. It is the same as saying if my little toe is touched, I can feel it and am conscious of what has happened and where it happened.

No Accident

The concept that I have just described should show that there are no accidents in Creation. If the Universe is that sensitive, how much more sensitive is the Creator? Personally, I believe that the movement of the flower would be felt throughout eternity because we are all a part of eternity and the Creation of God. God knows and feels my pain when I stub my toe or when my heart is broken. I am a small and insignificant thing as compared to eternity, but God can still feel my presence no matter what my circumstances are. Moreover, I can feel His presence no matter where I am or where I go. It is amazing that, of all the Universes that He has created and of all the marvelous things that are in them, God loves us (humans) more than any because He made man in His own image. We are the only creatures that He breathed His own breath into to make us live eternally. I am a part of Him, and He is a part of me, and we are all eternal beings who can never be destroyed but will experience either life or death eternally.

I watched a documentary on TV not long ago and found it to be somewhat disconcerting. Man has always adhered to the analogy that *nothing can happen in zero time*; please read that again. However, when talking about how the Big Bang created the Universe, their analogy was astoundingly close. They said that a very small amount of hydrogen, as small as an atom, suddenly and instantaneously erupted, and suddenly and instantaneously the Universe, which is billions of light-years in size, was born. One light-year = 5.8657 E12 miles or 10 trillion kilometers; i.e., the distance that light can travel in one year. This analogy is very complex and very perplexing to understand when you try to put a time element into the equation. For me, it is much more believable to think that God created the Universe and all that is in it within a period of six "God days," without using time, as man knows time, as a factor. To reiterate, and using my own analogy, the finger of God touched the eternal space at a particular point, and "BANG," it was created.

The disconcerting part concerning the scientific analogy is not that it happened instantaneously, but that man has changed his rules to make it fit. Folks, you can't change what is set in stone. I am stating that it is a fact here on this earth and in the Universe, as we know it, that *nothing can happen in zero time.* However, with God, all things are possible, and might I say probable, when it is done according to what He has said in His Word. That is what really confuses mankind; we think that God is bound to our laws of nature, and because of this, man tries to change his rules to prove that there is no God. If science is so perfect, why do they have to change any rules? This is very disconcerting to me because I have a lifetime wrapped up in trying to learn the rules and the laws of nature, and now they want me to unlearn them. Man, or science, cannot prove the Word of God to be true or false because it is an absolute value. I don't know if God used hydrogen as a catalyst to create the Universe or not, but if He did, then where did the hydrogen come from? Well, as I quoted earlier and science has recently proved, it came from nothing, and that is what man has proven so far: nothing.

Maybe when man finally learns that there is no time in eternity, he will realize that only God, not a Big Bang, created everything, including the nothingness and the juxtaposition of space. Science has always said that there has to be a reason for anything and everything; otherwise, it cannot exist. But now they are saying that "it just happened" or "it was an accident." I personally like the original hypothesis better: there has to be a reason for anything to exist. On this Earth and in the Universe, my friend, we must have laws to guard against impropriety or injustice of any kind. If we keep trying to change the laws and rules of nature, then it will lead to further confusion and disorder. God's Word says to *let God be true and every man a liar.* After centuries of denying the Word of God, they are concluding that what He says is true but saying it was man who proved it. This is absurd when we consider that God said it thousands of years ago. If I have stated, or will ever state, any untruths in my writings, it is because of my own ignorance, and I need to study more; nonetheless, *let God be true and every man a liar*; this means me as well.

And the earth was without form and void. Have you ever really concentrated on this statement? As the great Socrates, who was also human, once said: *"I am the wisest man alive because I know that I know nothing."* But certain phrases in the Bible intrigue me, and that is why I dwell on them for some amount of time. What does *without form and void* mean?

Form:

Manifestation: The particular way that something appears to be.

Basic structure: The shape or structure of a thing that gives it a distinctive character.

The shape or appearance of a thing that makes it identifiable.

Void:

Having no contents

A large empty space

Vacant, not occupied

Negating form takes away its shape or structure that gives it a distinctive character. In other words, the Universe was in chaos; entropy prevails, and everything was in disarray. It is only basic logic to think that it is going to take more than a magnetic field, time, and an accident to make order out of so much disorder. The most understanding we humans can comprehend, as pertaining to "VOID," is an empty jelly jar or a bounced check at the bank. Note that it has no contents; it is not occupied; it has no use or effective purpose; it is a large empty space.

Now, we have something that has no shape, contains nothing, and has no redeeming qualities whatsoever. So, what do we call it? Science has a name for it, and it is called "antimatter." Not long ago, if you had mentioned this to anyone, then you would have been placed into a straitjacket and taken to an insane asylum. But today, it is common knowledge to the science world: there is such a thing as antimatter, or perhaps "dark matter," or "negative matter," or whatever they are calling it today. It doesn't "matter." Who knows? But God said it existed in these first two verses of the Bible and that He created the Universe from nothing. I like to think of this as God taking a measurement of the Universe. I don't know if this is true or not, but it is as likely as anything science has surmised. Be that as it may, Christians are the wisest people alive because we know NOTHING!! We understand that if we take the dirt out of a hole, what is left is negative matter because the matter has been removed! Well, maybe I am being a little too sarcastic here, but the principle is the same. Scientists seem to have a problem understanding why there is much less antimatter than matter when it should be the same amount, according to them. I cannot get very scientific with this because I am not a scientist. But I have read about it to some degree and saw a symmetry to what I was taught in electronics. When an electron leaves an atom, it leaves a hole or a positively charged atom. But then, another electron will be attracted to this hole and fill the void, bringing the atom back to its equilibrium state. And so it is with the whole Universe; the symmetry is there. All we need to do is open our minds to it and not overthink it. The reason, I think, that the Universe has much less antimatter is that the atoms have reached a neutral point, or equilibrium, through the recombination of the electrons. Nonetheless, the Universe is still expanding, thus causing more entropy or disorder.

When we were defining form as it is implicated in Scripture, we stated that negating form takes away its shape or structure, which gives it a distinctive character. In other words, the universe was in disorder, and everything was in disarray. We also talked briefly about antimatter and stated that God was surveying this area of space, which was in disarray but in equilibrium because it was in eternity.

What did God do here, and what was He observing? What He saw, I think, was atoms and electrons that had no molecular structure and were loosely bound in eternity. I hope I'm not getting too technical, but I am explaining it the way God is revealing it to me through His Word.

Anyway, here is where God created water, H_2O, and then submerged the entire universe, or this area of empty space, with it. Water keeps heat to a minimal amount, so this was God's cooling system. Then He formed Earth and its molecular structure and placed Earth across the universe in an ordered fashion, as only God can do, while keeping it submerged in the envelope of water.

Now, you may be thinking, "Where did the atoms and electrons come from?" I have to say that I don't know. What I mean is that they could have been spoken into existence while God was observing, or they could have been in eternity with God from the beginning. Either way, I believe the molecular structure of objects, places, people, and things is the creation of God, for without it, nothing exists. Don't misunderstand; God could have chosen to do it another way, but this is the method that I believe He chose to glorify Himself, and I'm good with that.

And darkness was upon the face of the deep. God is going to great lengths to describe what He has done in detail. We cannot comprehend all the details, but they are there nonetheless. We see here that there was complete and total darkness in this area of empty, formless space, which God's finger had touched. Actually, I don't know if He used His finger or not; this was my own conceptual understanding. All God has to do is want it to happen, and with the very thought, it is done. We mortal, sinful beings are the ones who need some physical contact to make things happen, not God. I surmise that, for better understanding, we need to draw analogies and sometimes use metaphors to bring things into perspective and conceptual understanding.

In my own mind, this darkness indicates the point at which the Big Bang theory and I part ways. The reason for this is that never have I heard or seen an explosion without there also being fire, heat, and light woven into the equation. Even a heat storm that is looked at from a horizon has a flash of light, whether there is a sound heard or not. Also, for an explosion to occur, there must be an element of time involved and, in this case, a considerable amount of time. However, I believe that to have light, God must be present, and eternity is completely dark without Him, for He is light, and in Him is no darkness at all. This, my friend, is the instantaneous part and is also why in Hell there is no light: because there is no God. It is still eternity but without the very essence of life, God's light.

As a disclaimer, let me state that, like any theory, including this one, there have been updates to the Big Bang Theory over the years since I began this writing. This is understandable, for I have made changes to my own theory even recently, but the concepts for both theories remain unchanged.

When I was in the military, my MOS was "Heavy Equipment Operator." I worked and lived in a rock quarry while in Vietnam, and my primary job was demolition. I blew things up using dynamite and plastic explosives. This is not the way our Creator works. He is the Creator, or Builder, and not the Destroyer.

To get a true understanding of how the universe was formed, one must have the insight to look at the signature of the Author who created it. We will discuss this in broader detail a little later on.

I noticed a video recently that stated that many physicists were now quietly walking away from the Quantum Mechanics Theory. I will not opine on their reasoning, except to say that they said in the video that "it isn't that it is wrong, but that it may not be real." I surmise that they think the theory is trying to equate physical reality to the spiritual. This is only my assumption, not fact.

Black Hole

While my mind is struck in wonder at the creations of God, there is one area that I am drawn to continually, and I wish to write about it as much as my little mind can comprehend. The black hole is an area that still has scientists scratching their heads in disbelief. It is a place that is completely dark. No one alive has ever seen a black hole, but its existence is a scientific fact. The gravity is so great that nothing can escape. It has no light because God is not there. Its density is incomparable to anything known to man. If one person were observing another entering a black hole, they would observe a point at which the person would stop in time. Therefore, time stands still. There is no passing of time. All the laws of science, even Einstein's theory, are irrelevant and defied because his theory reaches a point of infinity or dividing by zero; quantum mechanics is the same. There is no mathematical principle that can explain the nature of a black hole. Since the gravity is so great, it causes great pressure and thus incredible and incomprehensible heat; it is at the center of our galaxy, not the center of our Earth.

Could this be the place we refer to as Hell? I am not at all placing doubt on the teaching that hell is in the center of our Earth. But maybe there is a difference between the hell we have always adhered to and the lake of fire spoken of in Revelation, which is also referred to as outer darkness.

If one enters a black hole, the material part of the body would be absorbed, stretched, and distorted within the black hole, although the body would be conscious. And just like a hologram, there would be a replica of that person's life or information forever and indelibly written on the event horizon of the black hole. The person entering a black hole would, for eternity, be looking at their own self and the life they have lived because, according to what the Bible has taught for thousands of years and science has recently proved, information cannot be destroyed. Wow! Man is really profound in his thinking, isn't he? Notwithstanding, science is in the process of learning the eternal existence of heaven and hell, but only to its own peril. They are correct in saying that no light can escape a black hole because there is no light to escape. God is not there. Any light entering a black hole, man's righteousness, is extinguished immediately and becomes darkness.

This, of course, is referring to the artificial light that God created, not the light of God, for He is light and contains no darkness, whereas a black hole is darkness and contains no light. It should really make a person realize and believe that there is a God of the Universe, but instead, the man who has proclaimed most of what I have said here, Stephen Hawking, has stated that there is no God. If he contends that there is no God, then why are some of the concepts that he has learned so closely related to biblical teaching? There is much evil within the heart of man. He will twist what God says into something entirely different and make it indiscernible.

In other words, Satan wants to hide the word of truth so that we cannot learn, and therefore we will become as brute beasts and slaves to Satan's trap. *(Proverbs 1:17) Surely in vain the net is spread in the sight of any bird.* Be careful not to fall into the trap. *Study to show yourself approved.* (2 Tim. 2:15)

A prime example of Satan's deception is the Big Bang theory. My friend, God has never blown things up to make things happen! Only Satan does that! More on that later, but for now, we will return to the black hole theory. Man has learned so much that has been plainly stated in the Word of God, and still, he declines the offer of eternal life and chooses eternal death instead. Inside a black hole, one will live and die at the same time forever because there is no passing of time, remember? Notwithstanding, we are all eternal beings and cannot die, but the fact that time stops within a black hole just emphasizes this even further. Throughout eternity, it will always be the present.

I have no trouble believing that we are in the latter days because the Word of God is being undeniably shown to the whole world in vivid detail. Also, this ought to really make a Christian stop and think because not only is information stored on the event horizon of a black hole, but it is also stored on the outer rim, or event horizon, of the Universe itself, according to science. Remember what God cautioned us about: every idle word will be accounted for? I realize that God does not need an event horizon or a computer to keep up with things, but it is man who needs the memory stick lest he forget the idle words. Be that as it may, it is a proven fact of the Bible that information cannot be destroyed, and science is now finding that out.

Sometime in the 1970s, I worked in a CB repair shop, and my boss told me a story that I would like to share with you now. He said that he used to work for an aerospace organization and monitored space for some time for the purpose of receiving audio transmissions from outer space. On one occasion, he remembered receiving some signals and began the process of cleaning the signal up to make sense of the transmission. When he finally got the transmission cleaned up, he listened as the amplified message blared across the speakers. It was a broadcast transmission of Fibber McGee and Molly that was being beamed back to Earth after at least seventy-five years of traveling through space and bouncing from one planet to the next. I have shared this story with you so that you will know that information cannot be destroyed. This is a stated fact in the Bible and is a proven fact of science. When your voice is impressed upon a radio wave of any kind, even your personal conversation between friends, it will make an indelible mark on the event horizon. Remember the flower moved by a gentle breeze?

Of course, God said that our sins would be cast as far as the east is from the west. However, I can't help but believe that unconfessed sin will be revealed, whether it is a spoken word or a thought process. They are both electrical signals and cannot be destroyed, but most importantly, God remembers.

And the Spirit of God moved upon the face of the waters. Here we are told two things: the Spirit of God moved, and there was water. Also, at this point in eternity, we cannot assume that the Spirit was moving or concentrating only upon our own planet and its local heaven or galaxy. Keep in mind that this is the Creation of the Universe, not only our own galaxy. The next verse implies that the light of God was shining at this point in eternity. Obviously, there was light only because God was present and moving! When God is present, the darkness will disperse. There can be no gray or dark area. The light in verses 3–5 is different from the artificial light that God created in verse fourteen, for it indicates the eternal presence of God and the distinct position and separation of His light from the complete darkness that was present before. So, how *"Big"* is God? God is a God of absolutes; either we are in complete darkness or complete light; either we are completely saved or completely lost. There is no gray area with God. On our Earth, we can experience some dark or gray areas of all things that touch our lives because of the pseudo, artificial light. But when talking about eternal things, it is either completely dark or completely light, and everything is absolute. Even the gold that paves the streets in heaven is crystal clear, pure, and absolute. This is why mankind cannot understand eternal things. It cannot be computed in any way, and all our formulas and mathematical functions will always be a number divided by zero. Who am I as compared to God? Science may be able to express this question mathematically, but neither they nor anyone else can understand it because God is infinite in His being as well as absolute in His creation. God = ∞, Man = 0.

When I consider thy heavens, the work of thy fingers, the moon and the stars, which thou hast ordained; what is man, that thou art mindful of him?

And the son of man, that thou visitest him?

For thou hast made him a little lower than the angels, and hast crowned him with glory and honour. Thou madest him to have dominion over the works of thy hands; thou hast put all things under his feet. (Psa. 8:3–6)

In Gen. 1:5, we see that God named the light and the darkness that He separated. He called the light *Day,* and the darkness He called *Night.* I have often wondered about this. Keep in mind that this is purely conjecture on my part, but why did God name light and darkness when the Bible states that God created the heavens and the earth and all that is in them? This is the first time that the Bible says that God put a name on something. Therefore, there has to be something special about the separation of light and darkness. I wonder if He wanted mankind to be ever aware of his eternal choices: *"Day"* being eternal life and *"Night"* being the eternal separation from *"Day," "light,"* and *"God."* Let me reiterate that there is an eternal separation between eternal light and eternal darkness. Also, as implied earlier, there is an undeniable symmetry throughout the Universe.

And the evening and the morning were the first day. The evening and the morning sound like God was working through the night, and maybe He was, but maybe He is already at the evening and is reflecting back over His day to where He started in the morning. I cannot comprehend that there was a morning or evening because of the absolute light or dark that is in eternity; well, that is God's business, not mine. Notwithstanding, please remember, there is no time as we know it when God was doing this work. Time, as man knows time, equals zero at this stage of the Creation process. Also, we don't have the sun, moon, and stars yet to reference a 24-hour period. But no matter what God is telling us here, our only responsibility is to believe and not make inferences to change or satisfy our own laws of science just because we do not understand His laws of science. We do not know how long this day was. As stated earlier, this is a work done within the realm of eternity, and there is no time as we know time here on planet Earth. What I mean is, if you were somehow placed on Mars, which is still a part of our own galaxy, time would change in that it would slow down. Therefore, one day is a matter of perspective, and we are talking about one day from God's perspective, not man's perspective. As stated earlier, God is absolute because God is! And He does not need or even have a reference to time. I surmise that if God wants it to be a millisecond or a million years, it is still within His infinite understanding, and He is in no hurry. It does not bother me that science dates back for millions or billions of years. All I know is that God is true and every man is a liar, and we cannot expect God to follow our little conception of time. If we do, we'll be dividing by zero, and we know the answer to that scenario. Also, I believe that God created a mature Universe as He did a mature man.

No matter what time your clock says it is right now, the Universe is still within the realm of eternity! God did create time, but He did it as a secular event, and we will say more about that later. Man invented the clock, and ever since, time has been confused and messed up. Man has said that one second is 9,192,631,770 vibrations of the cesium-133 atom as measured via an atomic beam clock. Man has also said that one meter is 1,650,763.73 wavelengths of the orange-red light from a krypton-86 lamp. I wonder if Superman would die if he were to come in contact with that lamp? I know that all this sounds ridiculous, and you are wondering where I am going with this, and I will be happy to tell you. Man has put a lot of work and mathematics into reaching conclusions of every type of measurement possible, and still, it is not perfect as God created it to be. He tries so hard to copy God and His creation and even calls himself a god. He makes statements such as *"Man has Created"* or *"Man is Creative,"* when God is the only one who can create, and He has made only one Creation, which is still going on because He is still working on me as well as His other marvelous works.

In order to create, it must come from NOTHING. Therefore, man always and forevermore falls short. Man has made great achievements and has learned recently that it is possible to travel faster than the speed of light, but with all his knowledge, he is still a failure. God does not need time to operate, for He works according to His eternal laws, and when He decides that the point in eternity is correct, He will say, *"COME UP HITHER!"* And it shall be done. As Einstein once said, *"We can do so much and know so little."* Einstein also said, *"If I have aspired to any new horizons, I have done so by standing on the shoulders of giants."* God is revealing to mankind the wonderful, awesome, and scary works of His hands. God is the giant whose shoulders we should stand on, not another man's shoulders. He said in His Word that the arm of flesh will fail you, and so it has been since the beginning of man.

I want to cover another concept of science, which was mentioned earlier but not explained in full detail.

Entropy

Entropy is an explanation or measurement of the disorder and chaos inherent in a system within the Universe. I will not address any formula's concerning entropy or thermodynamics. My objective here is to show that since the fall of Adam, the entire universe has been disordered and in chaos. If you have a bag of colored sand, you may be able to separate the many different granules and put order back to the disorder; but in the process, you would use more energy and thus cause more disorder. Therefore, though energy is conserved because it only changes states; i.e., your breakfast has been depleted but much has been accomplished, entropy increases, and thus the disorder will also increase*. With man's best efforts, he cannot cause an egg to spontaneously unscramble or cause a burned match to spontaneously re-form as it was before. These are what man refers to as irreversible changes*. It is the same way with the state of the world that we live in. Entropy prevails and increases, and only with the intervention of the God of the Universe will order be restored from so much disorder, as He did at the beginning, because the Universe and everything in it is headed toward maximum entropy, and it has been headed this way since the fall of Adam.

Jesus has revealed to us that He is the God of entropy. He replaced the ear that Peter cut off, fed a multitude of people from a few fish and two loaves of bread, and raised the dead, along with many other miracles, not to mention the enormous part He played in the Creation process. As a logically minded person, I am reluctant to call these things miracles because they didn't just happen. It was the power of God revealed. Scientists are now saying that these things were miracles, but they just happened; they were accidents; they were some magical, mystical events. Actually, it was God performing entropy control, and that is something man will never accomplish while on this earth.

I cannot believe how Satan is changing the meaning of so many words in our language; i.e., miracle means accident or a magical, mystical event from a worldly point of view, just like gay has been changed to something ugly and despicable. God said that if man did not give Him glory, even the rocks would rise and praise Him and magnify Him and worship Him. Since man will not read the Bible and believe it, or since he insists on changing its meaning, the Universe itself is crying out and shouting out to mankind, saying, *"I AM DESTROYING MYSELF!"* But he still won't listen because he has become like a brute beast that is incorrigible and stupid. He can do so much and know so little.

Before I go any further in this observation, I want to express my appreciation to those who have offered their insight and help. God gave us free will, and sometimes that free will gets distorted and in the way. Notwithstanding, when we think we know something, it causes many misunderstandings and more chaos, or more entropy. I like to refer to this distortion as dividing by zero because, if it doesn't fit with the Word of

God, then the computation or thought process itself will point to and give glory to God. Man = 0, God = ∞. We must be very wise because we know that we do not know everything. Thank you all, my friends.

And God said, Let there be a firmament in the midst of the waters, and let it divide the waters from the waters. I have a mental picture of verse six in that I am tempted to put a biblical poet's word at the end of the verse, which is *"Selah."* I need to be careful here because there is much to be considered, and the details are a little vague in understanding. First of all, how far in space are we talking about? I surmise that the first thing I need to do is find out what a firmament is.

Firmament: the sky, as considered as an arch (Microsoft Works dictionary). This definition is really too simplistic to explain what we are talking about. The firmament here is the one that reaches the third Heaven, where God dwells, and circumvents every galaxy and every planet in the Universe.

Therefore, this firmament is all of space as we can comprehend it. Some believe that this firmament is only the earth's atmosphere and the heaven is everything above that. However, I am convinced that the former explanation is more accurate. If it were only the earth's atmosphere, then it would limit God's creation. He said that He created the heaven and the earth, so it must encompass our Universe at least.

Scientists have stated that it all happened with a hydrogen atom; is this correct? Well, maybe. Water is made up of two molecules of hydrogen and one molecule of oxygen; so far, science has it half right, or maybe a third. Be that as it may, they are still scientifically wrong. God wants us to reason these things out logically and truthfully from His word. When I think back over my life as an electronics engineer and technician, I see that the objective of each project I worked on was to keep it from ever blowing up. That would make me look bad and unprofessional to the company and cause a loss of confidence among my associates. It is only when everything comes together smoothly that we can have real control over the machine. I believe that God has the same mindset concerning the Creation process. He didn't need to blow things up to make things happen. The difference between us is, He is perfect, and I, being human, will blow things up no matter how hard I try not to do so. I cannot help but make mistakes because I am man, but God cannot help but be perfect and mistake-free because He is God.

So, the first thing God did to create the Universe was to envelop it in water from the third Heaven to the end of creation.

And there was darkness upon the face of the deep. Why did He call it the face of the deep? To me, that indicates water. Therefore, if we have this crystal-clear, clean water and we submerged a handful of dirt into it, we would call that muddying the water, correct? I surmise that this is what God did when the Spirit moved. He troubled the water for the first time in history. Notwithstanding, it all started when

God created water, and everything else followed close behind. In electronics, I was taught that electrons cannot be manufactured nor destroyed. That is why man cannot create anything; he only uses what he sees on the horizon. Wouldn't it be a wonderful thing if all mankind would stand on the shoulders of God the Creator instead of the shoulders of flesh? If that were to be, then we would all be geniuses. The same concept concerning the electron is the same with all forms of energy, even man himself, because energy is the same no matter what, and God only made a certain amount of the stuff. It cannot be created anew or destroyed; it only changes states.

I suppose that by now, I have given away my analogy of creation, and that being that creation is indicative of the virgin birth when the water broke and a child was born. Most people don't compare this breaking or separation of water to an explosion, but I suppose that the woman feels as though it is when she is in labor.

Symmetry (Homogeneity)

The waters broke, and a Universe was born. This analogy confirms the sameness of Creation. A lady who has had an enormous impact on what we have learned concerning the symmetry or homogeneity, the sameness, of space and the Universe is Amalie "Emmy" Noether because she was the first to observe and explain the symmetry through Noether's Principle. She was born in 1882 and was an outstanding mathematician who did most of her work in abstract algebra. After a long struggle, she won the right, as a woman, to lecture without pay at Göttingen University in Germany. It was then, in 1918, that she presented the results of an analysis dealing with symmetry that became a guiding principle for contemporary physics. Noether taught at Göttingen until 1933, when she came to the United States after the Nazis learned that she was Jewish and expelled her from Germany. She passed away in 1935 but left behind this most beautiful and profound concept of science*.

If you are one who contends that there is only one Universe, that is okay. There is nothing unscriptural in thinking that there is only one Universe; after all, that is the one we are discussing. However, I like to think of eternity, the mother, as being endless in space and God the Father as being infinite in magnitude. Another thought is that God is the Great Creator and is still creating, even as Satan is the Great Destroyer and is still destroying. It is eternity that has given birth to many offspring, Universes. I am not trying to push my thinking down your throat. I am sure that you have your own thoughts, as I do, about issues such as these. I like to think of the planets and stars, or galaxies, as being the many organs that keep the body functioning properly. Therefore, the Universe is the child of a big God, and it was the first of many miraculous births which God has created. You see, God is the only one who can create, and He only created a certain number of electrons, atoms, and Universes, of which we are all an integral part.

None of these basic things can be created or destroyed by man or Satan, but they can be used and altered in a way that is detrimental to life as we know it and life as God intended. We are lulled into thinking that when man sinned in the Garden, it only affected him and his eternal soul, but this is not true; it affected the entire Universe because we are one with Creation and the Universe. This is the same as thinking that energy cannot be re-created or destroyed; it can only change states. Man's sin has changed the state of the Universe, and from that time until now, and until the Lord returns, the Universe will be locked into a state of destruction. Entropy prevails, but so does the symmetry of space.

And God made the firmament, and divided the waters which were under the firmament from the waters which were above the firmament: and it was so. This is another profound and wonderful statement that God has made as we progress through this miraculous birth of the Universe. Note that the verse says *"God made."* The creation has been completed, and the birth is now in progress. God is not creating here but delivering the child with expert love and care. Verse eight describes the view as seen from a father's eye. Like a proud father, He was so pleased, excited, and overjoyed, as any father would be, that He named the separation of the water *"heaven"* and marked the event on His eternal calendar. I can just imagine the Lord saying, *"On the morning and the evening of the second day, eternity gives birth."* Maybe this part of my analogy of creation is a little far-fetched, but I truly believe that the concepts are very accurate. Man tries to understand, in his own mind, the laws of nature as he interprets them concerning creation, but the true answers evade him because he is unable to create. We do not understand the miracle of birth or even how any child is formed with all the different organs and vessels that make up the body. We can only see that it is just the cellular structure that makes things happen in such a way that life is created here on Earth. This is why we refer to it as the miracle of birth. But so much of the miracle is missing to us and comes to light only when we believe in the God of the Creation. Of course, if we understood fully how the miracle is performed, it would no longer be considered a miracle. You see, we think that God works on the atomic level because that is as far down as our minds can comprehend. But to create something, God has reserved a level of understanding for Himself, and it is that level of understanding where God actually does His work. It is the infinite level of understanding. Einstein once said: *"There are two ways to view life. One is as if there are no miracles, and the other is as if everything is a miracle."* I believe this is a true statement, and I am one who believes that everything is a miracle because it all came from an all-knowing, infinite God, and it all came from nothing. We use that term very loosely because we do not understand the God who created us. We have not yet learned how to divide by zero; mathematically speaking: X/man (0) = (∞) God. Man is so stuck on himself that he thinks that nothing can be compared to him, but even mathematically, he still comes up short because zero will always be zero.

To the Work

And God said, Let the waters under the heaven be gathered together unto one place, and let the dry land appear: and it was so.

The waters under the heaven are where God started preparing the Universe in such a way that it can sustain life. It does not bother me that God may have also created other life forms on other planets throughout the Universe because the Universe is so vast that mankind will probably never find out all there is to know about it; in fact, it would take a lifetime of traveling at the speed of light to get to the outer rim of our own galaxy, much less the Universe.

Therefore, God chose not to mention these other life-supporting planets, and I am okay with that as well. However, let me say this: if God did create other beings, they will not have any more control over the Universe than we have. And they will not be some mechanical robots that are going to annihilate everyone on earth, for they will be created in the same image that we are. People today are really getting stupid with the notion that ancient aliens started wars and such with our ancestors. This thinking is absurd.

Our war is with principalities and the powers of hell, and it is Satan that is telling you that the Moon is an alien space station, and you should have more spiritual intellect than to believe that bunk. Okay, I am about to get on another book here, so I need to rein it in.

Getting back to the waters, this earth could not support all the water that was displaced in verse six all by itself, so that is why God said in verse nine, *"let the waters under the heaven be gathered together,"* rather than the waters which are upon the earth alone. This is speaking of all the waters that are on and around each significant planet and galaxy and in the Universe.

Now, with that said, let me make another point here. What would happen to a vessel of sorts, or a closed container, if it were taken to the bottom of the sea? It would more than likely be crushed by the tremendous pressure. Therefore, if our own galaxy was under enough water to reach to the third Heaven, I believe that this amount of pressure would form the planets and perhaps make a lot of diamonds and other carbon-based elements in the process. It would also account for the carbon dating that science adheres to as well.

I am not trying to prove any scientific fact here; I am only stating what the Word of God is showing me from the horizons as I attempt to stand on the shoulders of His word. So, dry land appeared upon each planet in the Universe. Also, according to verse ten, the dry land across the Universe was called earth. It doesn't matter if you are on The Third Rock, Mars, or Jupiter, the soil found on each is probably made of the same compounds which God named "earth." Every planet has its special purpose in the cosmos because God doesn't make junk. And I have trouble believing that God had to make the Universe and planets so vast and massive just to support one little life-supporting planet that we so proudly call Earth.

God did not call this planet "Earth," but rather all dirt on every planet and in every galaxy is Earth. Then man came, and with his man-made pride and self-righteousness, decided that everything God did was done on this planet, thus trying to limit God's Creation. Therefore, this must be the earth and the only planet in the Universe that God gave a name to.

Do you remember the scripture in Job 1:7, where God asked Satan, *"Whence comest thou?"* and he said, *"Going to and fro in the earth and from walking up and down in it"?* Does this shed more light on what Satan was doing? He was scouring the Universe, looking for something to dishonor God with.

Personally, I like the "third rock" as a better name for the earth we live on, and I am convinced in my own mind that if God had named the planets, this would have been as good a name as any. I do not believe that God intended to name any specific planet Earth, but rather, this is a generic name for anything that has dirt as its primary composition.

Back when Galileo was looking at the sun through his new telescope that he had just invented and was blinded by the same telescope, people thought that the earth was the center of the Universe and all other planets orbited around it. To believe anything different meant that you would lose your head, literally. People at that time had strange religious beliefs. That is to say, they worshiped the creature more than the Creator, so I can see how they might even believe that the third rock had preference and dominion over the others. I surmise that this is why they assumed that God was talking about this planet when He gave the name Earth in the Bible.

Notwithstanding, in chapter two, it is evident that He is referring to the third rock because of the geographical outlay; more on that later. Nonetheless, God called the dry land on all planets Earth, and the gathering together of the waters on all planets that apply, He called Seas: and God saw that it was good. In order to get a more vivid picture of Creation, one must understand that we are talking about eternity and the Universe, not just one planet.

Let the earth bring forth: verse eleven illustrates the work that the earth, including the third rock from the sun, was made to accomplish. God is almost finished with His creation and has now started allocating jobs that He expects of the Universe. *And the earth brought forth grass, and herb yielding seed after his kind, and the tree yielding fruit, whose seed was in itself, after his kind: and God saw that it was good. And the evening and the morning were the third day.*

Keep in mind that we are still referring to "earth" as being all dirt and rock on every planet. *(Gen. 1:14, 15) And God said, Let there be lights in the firmament of the heaven to divide the day from the night;* the lights spoken of here are across the Universe, and each galaxy has its own lights.

Here is where the first explosion, if there was one, occurred. I like to think of this as a father instilling enlightenment or knowledge into his son. After all, the Universe has a work to do, as we see later, and it will have to be conditioned to handle that job. *Give thanks and praise unto the Lord, for his mercy endures forever.*

We should read Psa. 136 before we go any further. II Cor. 9:15 says: *"Thanks be unto God for his unspeakable gift."* This, of course, is the gift of His only begotten Son, but the analogy here is so very similar to God's love that I could not help but state it in this manner.

You see, light is the word, and the word is truth, and truth is Jesus, and Jesus is God; can you see the symmetry? What we are reading in Genesis is a gift from a loving Father; the gift of knowledge, an exaltation of His majesty. And what we do with that knowledge can be handled with wisdom, or it can be handled foolishly. I believe this to be indicative of a father who is having that gracious, loving, long talk with his young son before he moves out on his own. *"Son, you have always had me around to instruct you, and now it is your turn to let your light shine."*

With this analogy, we can reference verse number two. There was darkness, but now I am giving you the light, wisdom, knowledge, and understanding, to light your way to life eternal. *"Ye are the light of the world. A city that is set on a hill cannot be hid."* (Matt. 5:14)

And let them be for signs, and for seasons, and for days, and years: and let them be for lights in the firmament of the heaven to give light upon the earth: and it was so. Right here is where man's understanding of time starts. This is where God sets the stars ablaze and the planets turning and rotating in perfect sync. So much so that if even one of them were to slow down or speed up at all, the effects would be catastrophic to every other planet in the Universe. They are all in a plane of perfection; they are in eternity.

You see, there are absolutely no accidents. An accident cannot make anything operate so perfectly. Everything does have a purpose, and there is order that came from disorder. Irreversible changes can be reversed! There is no way under the sun, fleshly speaking, that it could have happened, but it did happen; and science knows it happened but refuses to acknowledge the one and only Creator who made it happen.

God controls entropy. He can make the lame walk again and cause the blind to see, create the Universes from nothing, cause time to stop and go, and make everything be in perfect sync throughout this vast and wonderful, yet so very delicate, Universe; an accident? Absolutely not!

His ways are not our ways because He does not work on the atomic level; He works on the infinite level. We cannot understand that level of thinking and never will in this present world. We just have to accept it by faith, and God will reveal to us the understanding that He intends for us to have. He will reveal to us all the mathematical terminology that we need to understand every concept of nature and the Universe.

I do not believe that God would keep anything from us that we wish to learn about His wonderful work or Himself because it is He that is glorified from true science. However, as stated earlier, God works on the infinite level, and there are some absolute truths that we must accept by faith until someday, in this life or the next, God deems it to be the appropriate time to reveal them to us, and only then will we be able to make sense of a division by zero.

And God made two great lights; there is that word again: "God made." Creation is over, and now God is making things, and people, out of what is in that creation. In verse seventeen, we see that God set the stars, or suns, and moons in the firmament to give light upon the Universe. Can you see, from an eternal standpoint, that this is referring to all the planets and not just the third rock? *And the evening and the morning were the fourth day.*

Now we come to verse twenty of chapter one, and I must say that I respectfully disagree with the header of this verse in my Bible. It says, "The fifth day; the second creative act - animal life." I believe that there was only one creation in this Universe, and God did that one. You may think that I am splitting hairs here, but if I let it ride, there will be some understanding lost.

The Bible says here: *Let the waters bring forth....* It has always been assumed that this is talking about the waters on the third rock, but we are visualizing this from an eternal perspective; therefore, it seems to me that the entire

Universe is involved. I cannot be narrow-minded about this. God is much bigger than any of us can comprehend.

Let the waters bring forth abundantly is referring to not only the fish of the seas and fowl of the air, but also the creatures such as mosquitoes, bugs, and other insects in the Universe. *And God saw that it was good, and He blessed them, saying, Be fruitful and multiply,* and they have ever since. *And the evening and the morning were the fifth day.*

And God said, Let the earth bring forth the living creature after his kind, cattle, and creeping thing, and beast of the earth after his kind: and it was so. Although these things are a miracle and a part of the creation of God, we still have to conclude that the process is not a creative event because it is not coming from nothing anymore; it is coming from the earth or waters.

The living creature here is referring to worms, as well as mice, and the cattle are the cows and horses. The creeping thing could be turtles or spiders, or Chevrolets. Notwithstanding, the beast of the earth is most probably the wild animals, such as cougars and other Fords; and all of them, and we, are an integral part of the creation process, but we are not the creation, and the third rock is not the only earth in the Universe. *And God saw that it was good.*

Please do not condemn me for my pun on Chevys and Fords; I have to keep a smile on your face. But in retrospect, God's Creation produced the metal and gave man the knowledge and understanding to assemble it anyway, so it is all God and of the creation of God.

Man Is Made:

And God said, Let us make man in our own image, after our likeness: and let them have dominion over the fish of the sea, and over the fowl of the air, and over the cattle, and over all the earth, and over every creeping thing that creepeth upon the earth. So, God created man in his own image, in the image of God created he him; male and female created he them.

Let us make man in our own image means that we are made according to the way He envisioned us to be. In other words, to control and have dominion over the Universe. Can you imagine having the freedom to go anywhere you wish in the Universe and perhaps even the third Heaven and beyond?

If not for the blessed breath of God that was breathed into man in chapter two, and also the fact that we were formed by His own personal hands, we would be like any other beast even today. But the fact that God chose to make us Himself rather than have the earth do the work, and also the fact that He did breathe into us the breath of life, does put us in a higher plane of intelligence than the other creatures and gives us the right to have dominion over everything else in creation.

Which brings up another thought. Along with the prestige of being dominant, there is also the awesome responsibility of taking care of God's creatures. If we mistreat them, then we will suffer the consequences.

In verse twenty-six, we read that God "made" man. This means that He formed us from the dust of the earth. However, in verse twenty-seven, we read that God "created" man in His own image, which is a creative process in itself, in a manner of speaking, because it is the first time that God used His own breath to breathe life. But in another way of thinking, God has always had breath because God has always been; He just is, and He did use something, His Breath, to make this happen.

Therefore, man did not come from nothing and was made rather than created; as Einstein would say, everything is a miracle. The only reason that I can figure that the Bible says man was created is the fact that he is a very integral part of the Creation process because God made him, or formed him, from the earth but breathed into him the breath of life and did it all by His own hand.

The breath of God made Adam a living soul, and this is the first and only time this has occurred in the creation process. Therefore, Adam was made and created as well. The same with Eve as she was taken from Adam, she also became a living soul and had all the attributes of Adam.

The following scripture is reiterated, of course, but I want to keep it fresh on your mind, and it will make more sense if it is tied in with the same scripture in Hebrews chapter 2. I will touch on this somewhat forthcoming, but for now, I will only say that Adam, man, was not just another creature. God loved him more than anything in creation and gave him power over the entire Universe. Also, Adam was made to be perfect in every way.

What is man, that thou art mindful of him? Or the son of man, that thou visitest him? Thou madest him a little lower than the angels; thou crownedst him with glory and honour. Thou madest him to have dominion over the works of thy hands; thou hast put all things under his feet:

(Psa. 8:4, 5)

In verse thirty-one, God is looking back over His creation and says that it is "very" good. Our God is very proud of the work that He has done, but man has messed up everything in creation with his sin and set in motion what is ultimately going to be the end of the blessed Universe and life as God originally intended. It will be the end of our era, and a new heaven and a new earth will begin; entropy prevails, and *the evening and the morning were the sixth day.*

That is where we are right now on God's timetable, and on the seventh day, we will take our rest for one thousand years, and then we will once again have dominion over the Universe as Adam did then and as Jesus does now. In the new heaven and a new earth, we will become the sons of God and joint heirs with the Son. At that time, if we wish to live on this planet, we may. However, if we wish to visit Mars or have a cup of coffee and fellowship on Jupiter or the Andromeda galaxy, we can also do that with just a thought. Our freedom from sin will equip us with all the benefits that Adam had as ruler of the Universe.

I wish to address more clearly some questions as to why God did this great and magnificent work of creation. I have given most of the detail on how God did it as perceived in my own mind because that is the main topic of discussion in this study. But for more understanding, I must include, with more detail, the whys of creation, and that in itself could become a book.

Think of the song titled "The Love of God." *Could we with ink the ocean fill, and were the skies of parchment made; were every stalk on earth a quill, and every man a scribe by trade, to write the love of God above would drain the oceans dry; nor could the scroll contain the whole though stretched from sky to sky.* I believe this song to be very valid and true.

There is no way that we as mankind can do justice to the glory of God, no matter how we try, because He is infinite, and my feeble attempts pale in effort to ascribe to His splendor and glory. Be that as it may, we can still formulate our own inferences and opinions from scripture and from what we see in nature and other scientific facts of nature.

I hope that I have indicated in my writings so far that the Word of God IS our one and only source of absolute truth, but everything that touches our lives and everything we learn of the Universe is a direct revelation of the God of the Universe because He created it to glorify Himself, and the whole of creation takes on a connotation or a likeness of God Himself.

Questions?

God so loves the Universe that there isn't anything that He would not do for it. Why did God create this planet different from any other planet that we know of today? I am not sure that He did create all of them differently in His original plan. I am with the thought that each one had a specific purpose in the Universe, similar to what the Third Rock does at this time: to glorify the Most High God. Nonetheless, as I said at the beginning of this article, if even a flower is moved by the gentle breeze, it is felt throughout the Universe. Man's sin has therefore affected the Universe. I cannot give scripture and verse for it, but my assumption is built upon what my own eyes see in creation, as well as Biblical truths, and I believe that even the rocks on every planet and the Universe itself will magnify and give glory to God. By the same token, I believe that the sin of man will also be shown and magnified throughout the Universe, because it is the mistake of man and not the mistake of God that caused entropy to prevail.

So then, God created the Universe and everything in it to be perfect and filled with His perfect light; then man brought darkness back into the equation with his sin, and ever since, the Universe has been destroying itself. There cannot be eternal life without the eternal light of God. As I stated earlier, I cannot see that God needed the Universe to be so massive only to support one little life-giving planet. Consider what we do know and can prove by the Bible concerning the mind of God. The Bible teaches that after the Great Tribulation, God will have a remnant of His chosen people.

Well, using that same thinking on a larger scale, perhaps the Third Rock is the remnant of the once abundant Universe of humanity. Can you imagine, in your mind's eye, that when Adam was in the Garden, he had the freedom to move from one planet to another and even up to the third Heaven instantly, because God gave him this power? The only man who has this power now is the second Adam, Jesus Christ. Everything that is done on Earth or in Heaven has a reason, and I am offering the only logical reasoning that I can comprehend.

Why did God create mankind when He had other intelligent creations? The Bible teaches that Adam was an eternal creature who had the very breath of God breathed into him, and thus, he became a living soul. Then God walked with Adam in the cool of the day. God loved Adam as His own son because he contained the very breath of God. Then, when Adam sinned, that really messed things up, and death passed upon all men, for that all have sinned. I have underlined the word all here because the action is all-inclusive of the Universe itself. Man is the most important part of the Universe, and therefore, what he does will affect the entire Universe. As I indicated earlier, if I were to hit my little toe with a hammer, it would affect my entire body, and the effects would be pronounced throughout my body, and so it is here. Angels are not a creation of or within our Universe. They are Heavenly creatures. Our Universe is created as a unit of God's body, just as the heart or lungs are a unit of our own body. Okay, the Creation is complete, and if you will notice in chapter 1 and verse twenty-five, God said: *"25 And God made the beast of the earth after his kind, and cattle after their kind, and every thing that creepeth upon the earth after his kind: and God saw that it was good."*

Now, we are still thinking about the Universe here and not just the Third Rock. Please don't think that I am questioning anybody's intelligence by reiterating my points; we are so accustomed to thinking from an earthly standpoint that it is difficult to imagine the concept of the Universe also being in God's plan, and we must often be reminded. Notice that I underlined the word *"Made."* This is saying that all the creatures and beasts throughout the Universe were made by God, and they were made to help man in the caretaking of the Universe. This includes the beasts that walk uprightly that archaeologists are so enthralled with today. Keep in mind that even though time has been initiated, as in verse fourteen, the Universe is still in an eternal life environment. Therefore, time does not have any aging effects, for there is no passing of time or reference to time until after man sins in chapter three.

"26 And God said, Let us make man in our image, after our likeness: and let them have dominion over the fish of the sea, and over the fowl of the air, and over the cattle, and over all the earth, and over every creeping thing that creepeth upon the earth." In this verse, we have two words underlined, which are *"Make"* and *"Them."* Now look, I agree that Adam was the one human form that God created, or in other words, breathed the breath of life into. But it is evident from here that Adam was not the only

"Man" that God *"Made."* This should tell you what the missing link is. It can be nothing other than the very breath of God. That is the only thing that separates us from any other manlike creature that they are digging up at this time around the globe. Now, something else to blow your mind: since Adam had the very breath of God breathed into him, it stands to reason that Adam was the same as God, only in the flesh. Please don't write me off as yet. We call Jesus the second Adam, right? Okay, Jesus said that *"All power is given me of the Father,"* which means that He is the Son of God and is God, right? Okay, when we see Him, we shall be like Him because we will be called the sons of God, right?

This same logic held true with Adam; he was given dominion or power over the entire Universe, and in that respect, we can consider him as the same as God, not equal to God, but of the same mind or essence. Now let me interject one other thing before moving on. In chapter one, verse number twenty, we find that Adam was naming every creature that had been made. But notice that there was not a help meet that was good enough for him to be found. This is because of the breath of God and nothing else. The help meet here is obviously speaking of a soul mate because Adam had plenty of help, company, and friends. Therefore, this mate had to come from Adam himself, just like God Himself had to be the perfect sacrifice on Calvary for the propitiation of our sins. The real problem with Adam came into play when he went against the command of God. That fruit was the only thing that God reserved for Himself, and Adam was assuming himself to be equal or superior to God when he took of it, as did Lucifer when he assumed himself to be equal or superior to God. The only difference is that Adam was given a second chance, and that is only because of the Breath of God, which will always be preserved.

As far as these other manlike creatures who did not have the breath of God, they were all destroyed, and the entire Universe was doomed. This was when entropy came to be, and Adam and Eve, as well as all other creatures, were locked or imprisoned into the Third Rock's environment. I stated before that earth is on every planet throughout the Universe, but now the only earth that man can imagine is the prison of this one planet, so he named it Earth. The sin Adam committed also started the passing of time that we are accustomed to today. So, did Satan set into play the entropy or destruction of our Universe, or did God destroy it? I believe that Satan did because he is called the destroyer. But Satan is limited as to what he can destroy because of the divine intervention or permission of God Himself. You see, it was Satan's objective all along to destroy God and all that He had made, but he cannot destroy us or anything without the direct consent of God. One may make the argument that Adam did not die immediately after he sinned. Well, he didn't die immediately in body, but he did die immediately by being separated from God, spiritually, and he also started to age, and that is when I think time as we know it started to be counted against mankind. Like I said earlier, time started back in 1:14, but it was only a secular event and didn't mean anything until Adam sinned. Notwithstanding, the sentence was passed immediately, and the chain reaction began. Even a flower moved by a gentle breeze is felt throughout the Universe. Therefore, we of today and every person since Adam have been born spiritually dead, and it is only Jesus Christ who can quicken us or make us alive again with His indwelling Spirit. Some may wonder about that statement, in that many babes also die both inside and outside the womb. The children are safe in Christ until they reach the age of accountability. In other words, *(James 4:17) "Therefore to him that knoweth to do good, and doeth it not, to him it is sin."* Also, *(Romans 5:13) "For until the law sin was in the world: but sin is not imputed when there is no law."* God does not punish the innocent, so that is why I say that the children are safe in Christ.

It took the very blood of God Himself to cleanse man of his sin because if there is sin in God, then there is no true God. As stated earlier, God is absolute in His Creation and infinite in His being; therefore, in Him is no darkness at all, and our father Adam was made to be the very image of God in every way. Therefore, if he sins, he will continually die for eternity. However, God's Love, Mercy, and Grace still abound for this Universe, and it is the blood of Jesus Christ, The Great I Am, The God of the Universe, that has given us the ability to once again become sons of God and joint heirs with the Son. Jesus said that ALL power is given to Him of the Father. Therefore, He is now sitting at the right hand of the throne of the Most High God. He's the only man who has fully accomplished that so far, but we will also have another chance ourselves, thanks to Calvary. He is a God of second chances, PRAISE GOD!!

The Seventh Day

As we continue in chapter two, we can see that the word THUS is the first word in the chapter. I have heard some writers indicate or infer that there were two creations, and one of them didn't make it, so God created another. That is a very ignorant analogy and indicates that God made a mistake, which is absolutely absurd. I have already stated the only reasonable explanation for carbon dating that I can think of. It is simply the result of the tremendous pressure that the Universe was under from displacing so much water at birth. To suddenly change from one extreme to another must have created a tremendous vacuum, I would assume, and thus, a lot of heat as well. However, we know from what we have learned so far that there is no time element in eternity or in creation unless you wish to include Genesis 1:14 and onward as part of the original birth of the Universe, and I do not, because the Universe was created from nothing, and everything else, including man, was *"Made"* by God from the elements of the creation. Therefore, this first word of chapter two ties together the first two chapters of Genesis, which explains in more vivid detail the *"whys"* and *"hows"* of creation.

The word THUS is an adverb and indicates a connection or affirmation of what has previously been said; that is to say, *"consequently,"* as pertaining to a previous statement. Yes, everything is a miracle, but there is only one creation, and that is why I split hairs on these issues. Everything must fit together, or else it will blow up in our face. For more clarity, we cannot assume that God's ways are arbitrarily man's ways because Man = 0 and God = ∞. *"God Created from Nothing"* and *"God Made from the Creation."* Therefore, everything is a part of creation, but there is only one creation. *"THUS, the heavens and the earth were finished, and all the host of them."*

"And on the seventh day God ended his work which he had made; and he rested on the seventh day from all his work which he had made." Here in verse two of chapter two, I am tempted to use the old but true cliché that states, *"History Repeats Itself."* If you really think about it, God made the nature of everything to have a sameness; that is to say that God Himself did the entire work in six days, and on the seventh day, He took His rest from the work that He had done. Since the Universe and man were created by God and in the image of God, there has to be a sameness throughout the Universe that has a cycle of a set pattern that will not be interrupted. This holds true for all things, and it holds true in spite of man's sin, which obviously had its impact upon the entire Universe. When man sinned, it was the same as the entire Universe sinning because man is one with the Universe. I firmly believe that when God created the Universe, He made all the planets beautiful and livable. But the destruction and desolation came when Adam sinned. This may seem like a strange teaching, but it is what I believe to be true. If you will note, God said that death passed upon all men, for that all have sinned. This not only includes man but everything

that God created within the realm of the Universe. Animals have a life cycle, as well as men, and so does everything else in nature.

For more clarity of thought, I wish now to risk the possibility of boring my readers for just a few paragraphs. Please keep in mind that God made or created everything, and this includes true science and the mathematics that explains it. Therefore, don't limit God or His power to only the depth of your own mind, but rather expand your own understanding so that you can comprehend, in more explicit detail, the awesome, magnificent, and incomprehensible power of the God of the Universe. Remember Proverbs 4:7: *"Wisdom is the principal thing; therefore, get wisdom: and with all thy getting get understanding."* And then Proverbs 1:7: *"The fear of the Lord is the beginning of knowledge: but fools despise wisdom and instruction."* Wisdom is the principal thing, but it is God who initiates it. Consider in your mind, if you will, how the planets spin and orbit around the sun, and how the moons spin and orbit around the planets, and they all make up the galaxies, which spin and orbit around each other and ultimately also the Universe. Well, with that thought in mind, now consider the molecular structure of any particular object, person, or thing. You will find that the only difference between any of these is its molecular structure. A copper atom, for example, has only one free electron in its outer valence, whereas a block of wood has none. Therefore, it is the way that these atoms are made, as well as the way they are arranged, that dictates whether you are holding a copper wire or a block of wood. The orbit of the atoms around other atoms, with electrons orbiting the nucleus of the atom and so forth, has a sameness with the Universe that God created.

Now consider the growth rate of everything that God created. You will find that everything in nature grows and decays, lives, ages, and dies at the same rate. This rate is well known as the natural logarithm, which is located on any scientific calculator and has a value of 2.718281828. I know what you are thinking; it is too much information for me as well, but I did not create the sameness, God did. And all man has done is observe what God reveals from the horizons. To further bring this thought of the sameness of the Universe into perspective, I will go one step further. In electronics, there are two devices that are called reactive devices; they are the capacitor and the inductor. The capacitor is voltage-dependent, and the inductor is current-dependent. The charge and discharge rate on both of these components is at the same rate, which is the natural logarithm. It has always struck me in amazement that two seemingly inanimate objects grow and decay, charge and discharge, live and die at the same rate that every living thing does. Folks, the sin of man affects all things, and nothing is exempt from its death grip. When I was in college, my physics professor would not allow us to use the Pythagorean theorem when making mathematical calculations. We always had to do it the hard way, and so we asked why.

The professor replied with the following statement: *"The Pythagorean Theorem is only a theory and not a law; notwithstanding, it has never failed yet, but the potential is still there. It is only a theory, and our calculations in this class shall be based upon facts and not theory."* This is very profound if you think about it. Remember Einstein's theory of relativity? Remember when we tried to explain the black hole, how his theory met its untimely demise, facetiously speaking, as well as the quantum mechanics theory? For many, many years, Newton's theories were a standard; but now they fall short when given certain conditions, and that is where quantum mechanics takes over. And so it will always be until someday, God answers all our questions, but for now, we will just be awestruck by the splendor of it all and the sameness of it all. We can also rest assured that the Holy Book of God is the one and only absolute truth, and everything else, including what you are reading now, is theory.

So, why have I gone to such lengths to show that there is a sameness throughout the Universe?

What does it prove? It shows that there can be only one Author or Creator and personifies Him as the only God of the Universe. Remember, when we quoted the Bible earlier from Luke 19:40: *"And he answered and said unto them, I tell you that, if these should hold their peace, the stones would immediately cry out."* Okay, so I didn't quote it exactly right before, but it is still true that the stones of the Universe will glorify our Lord and Savior, Jesus Christ, and the homogeneity of the

Universe and everything in it only glorifies and personifies Him more.

Let's look at a personal level. People have a unique way of doing what they do, from tying shoes to flying jets or from designing machines to sweeping the floor. I used to be able to look at a specific circuit within a machine that was designed in our plant and tell you which technician did which work because they were all different with the same work but were identified by a sameness that related only to them. In other words, they left their signature on the work. Not an actual signature, but the work identified them as the tech who did the work. The forming of our Universe and everything in it absolutely could not have been an accident because everything has the same design. And it was absolutely created by only one, and that one had to be the God of the Universe. How do I know? The signature is the personification of God Himself. You may say that there are three in the Godhead, and this is true, and they all three abide in one. Although it would be commendable to read the following verses through verse seven, I feel as though I have said as much as God has given me liberty to articulate concerning them, and therefore, I should pick back up in verse number eight.

"And the Lord God planted a garden eastward in Eden; and there he put the man whom he had formed." I mentioned before that there was a place where God was relating only to the Third Rock, and we are now there. Now, let your mind sink in thought as to the location and geographical lay of the land. No one

can pinpoint exactly where the Garden of Eden was or is. We are only told in the Bible that it is where four rivers head. The Euphrates River is definitely one of them, and the Tigris River is another possibility. The other two rivers are a mystery, but one assumption, and perhaps a very logical one, is listed below as to the modern-day location of the Garden of Eden.

The Genesis creation narrative relates the geographical location of both Eden and the garden to four rivers: Pishon, Gihon, Tigris, and Euphrates, and three regions: Havilah, Assyria, and Kush. While the Tigris and Euphrates rivers are well known, the Pishon and Gihon are not known at all. Gihon is sometimes identified with the Karun River of Iran that flows into the Persian Gulf below Basra. Kush is sometimes associated with Egyptian Kush, but it is also associated with the Kossaei, the homeland of the Kassites, Kashshu, in the Zagros Mountains.

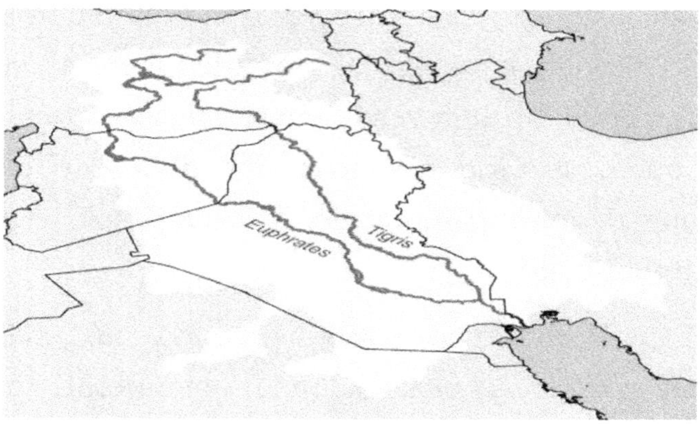

The Garden, they say, is under the water of the Persian Gulf. I cannot argue with that, but it does seem strange that if this river fed four other great rivers, then Eden would logically have been an enormous place because this one gigantic river watered it. I would rather consider this as one of many assumptions and continue with the concept that Eden, wherever it was or is, is a very, very large parcel of land.

God made woman from the rib of a man to be a helpmate to him. However, like everything else man messes with, he has placed an evil, selfish, and prideful meaning on the relationship. A helpmate does not mean a slave. It means that he should always give audience to her, for she does exude a great amount of intelligence and can be of great help to keep us out of trouble and confusion if we are willing to listen. My pastor is continually making a quote when he wants to emphasize a statement: *"Let Me Help You With This"* or *"Can I Help You?"* That is a wonderful saying and very profound. That is exactly what God created woman to do, be a helpmate. It is not only man, but also the woman, who is responsible to present information or instruction. It is up to the man to learn, compile, and implement the knowledge of God, even as it is the woman's responsibility to do the same. We should have a sharing, giving relationship with all believers. It is a very sad thing and a waste of brain power when we let pride get in the way of |TRUTH|.

There are a few laws of nature that we need to familiarize ourselves with to better understand what God is teaching us about his Creation. Keep in mind that a law is a proven fact. This does not mean that man made the law, but rather observed what God has done in the Creation process. A law, whether it is a mathematical law or any law of God, is an inherent statement that cannot fail. As stated earlier, it must reference |Truth|. Also, as a disclaimer, these laws apply only to man and the Universe because God is the Creator of the Universe and the symmetry and homogeneity thereof. We usually equate Matt. 5:18 to the Old Testament law, but I wonder if this cannot also mean any of God's laws, including the ones listed here.

Scientific Laws to Remember:

1. Nothing can happen in zero time. If this could happen, then dividing by zero would be a cinch, and we know the answer to that: Man = 0 and God = ∞.
2. There must be a reason for anything and everything; otherwise, it cannot exist.
3. The Word of God is the only absolute truth.
4. There is a sameness in nature from the atomic level and throughout the Universe.
5. All theories must reference absolute truth in order to maintain validity.
6. Entropy prevails, and our Lord is the God of entropy because he is the only one who can create.
7. God is infinite in his being and absolute in his creation.
8. Information cannot be destroyed, whether it is an audible word, a thought, or transmitted through the airwaves. They are all electrical waves.
9. Electrons or atoms cannot be recreated or destroyed because the Creation has already been completed.
10. All matter, including man, is made of the same stuff: the atomic structure. Laws 8, 9, 10, and 11 prove Law 4.
11. Energy cannot be destroyed but is rather conserved because it only changes states. Some folks are very

confused about this, as well as other laws of nature. They try to mystify and glorify the laws of nature. They worship the creature, energy, rather than the Creator, God.

12. Entropy prevails, where entropy is the measure of the disorder in a system within the Universe.

The Big Bang theory is a lie of Satan. Just saying that does not prove it, but it does make me feel better to say it first. As far as proof, let me say that only Satan would blow something up in order to accomplish his goal. His sole purpose is to destroy and cause the destruction of God's Creation. God doesn't want to blow things up and never has, because he operates on the infinite level, and Satan can't do that. Satan cannot do anything in zero time as God can do. You see, Satan says that a nuclear explosion is what created the Universe. That is impossible because it takes time to explode an atom bomb or perform any work inside the bounds of the Universe. We know for a fact that there is no passing of time within the realm of eternity, so there was no explosion; it was God performing entropy control. I have offered only one theory concerning creation; therefore, being based on Biblical principles and the laws of God, it is very close to absolute truth that will not fail, though the potential is there only because of man's intervention.

This is all that I have been led to write concerning the creation of the Universe. For me to continue would be useless and redundant, for I have written that which is in my heart. For now, I believe that I have completed the task that our Lord has impressed me to compile to the best of my knowledge and ability. I am sure that from what I have said already, the reader can see that whatever is discovered by science to be true and factual is also stated by God from the very beginning of time until the present day. Science is not wrong on many of its discoveries, but its inferences or theories are far off-key if it tries to exclude God and Biblical principles from the analogy. I have given my own theory in these writings, and it is the same as anyone else's theory, except that I have tried to implement an understanding of the truth of God into each statement that I have made. I do not know if my illustration of creation is 100% correct or not, but it is much more believable than any Big Bang accident that has been proposed in the past. I will always believe that we are one with the Universe because of the sameness of Creation from the atomic level to the infinite level of understanding. And someday, we will be one with God thanks to the work done on Calvary by our Lord and Savior, Jesus Christ.

One of my favorite earthly beings was Nikola Tesla. In my humble opinion, he was the greatest inventor of all time, and I would like to close with one of his quotes: *"The scientists of today think deeply instead of clearly. One must be sane to think clearly, but one can think deeply and be quite insane."* - Nikola Tesla

Symmetry

Physics state of invariance: a state of invariance shown by some phenomena when changes of orientation, charge, or parity are made.

Homogeneity

Quality of being the same: the quality of being of the same or a similar nature.

Quality of being uniform: the quality of having a uniform appearance or composition.

Cited Work:

Physics; Eugene Hecht; Adelphi University; Pages 8, 116; Brooks/Cole Publishing; Pacific Grove, CA.

http://en.wikipedia.org/wiki/Garden_of_Eden

www.ingramcontent.com/pod-product-compliance
Lightning Source LLC
Chambersburg PA
CBHW070646240926
51846CB00003BA/802
* 9 7 8 1 9 7 2 7 5 0 1 0 0 *